The Natural Path

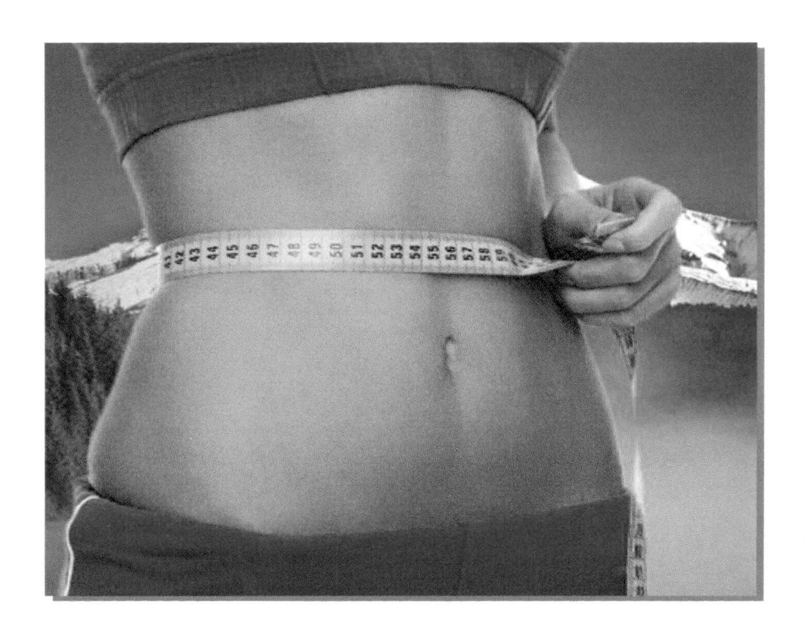

HCG DIET & WEIGHT LOSS GUIDE

A MEDICALLY SUPERVISED WEIGHT LOSS PROGRAM

Designed and Facilitated by:

DR. ROBERT MADDA N.D. &
DR. CARA PHILLIPO-MADDA N.D.
NATUROPATHIC PHYSICIANS

table of contents

Jump-start your metabolism and your life with the medically researched

Natural HCG WEIGHT LOSS PROGRAM … that really works!

Finally, after many years of research, we have developed a weight loss plan that is safe, effective, long lasting, medically researched and fulfills our commitment to higher health. This medically supervised HCG Diet program is based upon over 60 years of research and clinical trials. In our own clinic, we have seen a consistent and positive outcome in most patients. I leave the office on a daily basis feeling inspired by the results that my patients have with this weight loss program.

The key to success is a combination of the HCG Diet menu plan, daily injections of pharmaceutical grade Human Chorionic Gonadotrophin (HCG) & a few, simple lifestyle changes. HCG is a hormone naturally produced by the placenta in early pregnancy. In the 1950's, a brilliant British medical doctor named Dr. Simeons discovered that HCG had a direct relationship with the mobilization of abnormal fat stores in the body. Dr. Simeons theorized that a therapeutic dose or prescription HCG, along-side a specific diet worked by calibrating the hypothalamus in the brain.

The HCG Diet was found to work as safely and effectively in males as it does in females.

In addition to daily injections of HCG, the program requires that you follow a specific very low calorie diet (VLCD) for 3 to 6 weeks. On this diet, most patients lose an average of ½ to 1 lb. per day (depends upon what you have to lose). Once the HCG injections and VLCD are complete, there is a 3 week "stabilization phase" followed by the "maintenance phase" where additional foods are slowly added back in.

It is as simple as that! By the time the maintenance phase is reached, we hope you have learned a new way of eating that optimizes weight management. Best of all, you will most likely be feeling better than ever!

*Just a reminder that the FDA does not approve the use of HCG for weight loss. Therefore, we use the HCG for weight loss (along with the specific HCG Diet) as an off label use of the medication.

Let the HCG Diet be a Springboard to Long Term Success with Weight Management

What we love most about facilitating the HCG Diet here at The Natural Path are the long-term results that most of our patients experience after adhering to the protocol. The results inspire us on a daily basis. **The HCG Diet is a full immersion approach!** It gives you an incredible opportunity to change many habits and patterns in your life while deeply cleansing your body and shedding unwanted fat stores.

1. **Bye bye food addictions.** The HCG Diet allows the body to let go of sugar cravings, processed food cravings and to actually enjoy eating healthy foods that nourish your body! This is worth celebrating!
2. **The HCG Diet can have a positive impact on hormones.** Many of our HCG Diet patients report better sleep, improved moods and energy levels during and after the program.
3. **Hello to new lifestyle habits.** The HCG Diet can help you learn to eat mindfully. Perhaps you will no longer eat until you feel uncomfortable. You will hopefully become more mindful of the nutritional value in food and eat to nourish your body! You will most likely begin a new exercise program too ... come phase 3!
4. **The HCG Diet can help you to feel good about eating real food!** There is no reason to be afraid of eating "fat" or any food that is dense with nutrition. This program will help you to establish a positive relationship with food in your life.
5. **The HCG Diet provides an opportunity** to establish new patterns in your life on many levels. You have an opportunity to springboard into a leaner body, a healthy eating regimen, a non-negotiable exercise plan (phase 3), and a YES to being your best in life.

We have seen the HCG Diet improve HEALTH!

If you are overweight, then you are at risk for developing any of the great American diseases of aging such as heart disease, diabetes, hypertension, stroke, hormone imbalances or fatty liver disease AND you are an excellent candidate for this program. **One of the things we love about this weight loss method is that it gets you out of the active health risk category very quickly and puts you on the road to optimal health without all the excess weight.**

NATURAL HCG DIET & BODY COMPOSITION PROGRAM

What this program includes ...

- Step by Step HCG DIET GUIDE outlining the phases of this program
- Comprehensive medical office visit/consultation before, during and after the program. Body composition analysis, lab work (if needed) and overall weight loss goals will be reviewed at this time.
- Weekly or bimonthly 15-30 min check-ins with doctor to discuss progress, obstacles and concerns. Medical guidance is available through the program via phone or email between office visits. Yes, we encourage our patients ask any questions they may have during the program. We like to be of support to our patients.
- Rx HCG daily injections will be supplied each week of the program
- A specialized "Slim Shot Rx" injection with B-complex, B-12, Carnitine & lipotropic nutrients will be given in our office with each medical HCG Diet follow up appointment. (optional $40)

Benefits of this Natural HCG Diet program have been known to be ...

- A chance to lose weight ~ increase energy, vitality & mental clarity
- Learn about the physiology behind "fat burning" and "fat storing" and how to change it
- Relief from chronic health issues including low energy, high cholesterol & high blood pressure*
- Enables you to transform your body in less than thirty days RESULTS WILL VARY FROM PERSON TO PERSON
- Targets the metabolic issues that prevent you from losing weight
- **Optimizes blood sugar and thereby sensitizes insulin to promote fat burning instead of fat storing**.

Results seen by many on the program ...

- **Weight Loss on average of .5 - 1 lb. per day that spares lean muscle and decreases hunger**
- A cleaner, leaner, more vibrant and energetic body
- We find that 75% of people do not gain more than 5 lbs of the weight back; those who gain the weight back often know exactly why this happened.
- Triggers and "re-sets" the hypothalamus gland in the brain to release stored fat
- Improve your overall sense of health, wellness and vitality.

PERSONAL STATE OF READINESS BEFORE TAKING ON the Natural HCG Diet

It is important to ask yourself these questions before undertaking this commitment

- **Can you commit to staying on the program for the time needed?**
- **How easily can you fit the program into your lifestyle?**
- **What opposition will you have from family and friends?**
- **What support will you have from your family and friends?**
- **Do you understand that this program needs to be followed exactly as outlined in this book in order to obtain optimal results?**
- **Do you have the time needed to complete a full round?**
- **Are you ready to change your life?**

MAKE THE COMMITMENT

We make many commitments in our lives. We make commitments to our family, school, job, loved ones ~ just about everyone else but ourselves. Well, it is time to make the most important commitment – the commitment to YOU. Feel free to start this program when YOU are ready. It may take a few days to prepare yourself. Clean the junk food out of your house. Get your mind set and Find true motivation within yourself to give this all you have got. Are you ready to shift your physiology and maintain a very healthy lifestyle? **Create an image of yourself living in optimal health**. This is your personal wellness commitment.

"I __________________________, promise to put forth my best efforts to follow the components of the Natural HCG DIET & BODY COMPOSITION program. I understand this program is fully effective only if I follow the guidelines (and do not cheat). **My date for starting this program to reboot my metabolism, create optimal wellness and become even more vibrant is ______________."**

Write down why you would like to do this HCG DIET & BODY COMPOSITION PROGRAM and live a healthier lifestyle.

Make a list of all the challenges you may face and how you will overcome them.

Potential challenge | Plan to Overcome

Reward yourself for a job well done!

For example, every week that I complete on this program I will reward myself by ...

Signed, ____

YOU CAN DO THIS!!!

BELIEVE IN YOURSELF!

We believe in YOU!

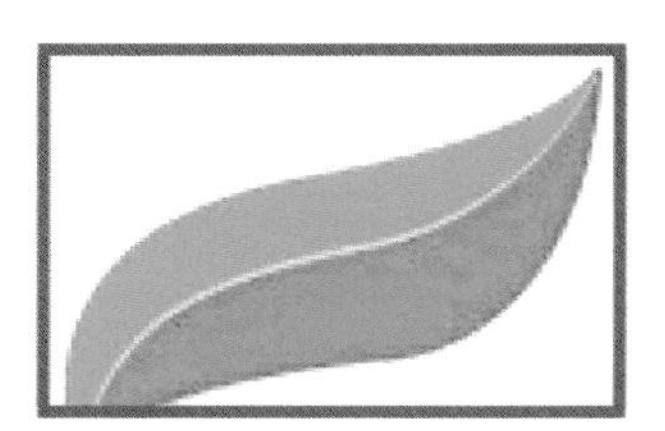

Commonly asked questions about The NATURAL HCG DIET

What is the Natural HCG Diet Protocol?

The Natural Path HCG diet is a unique diet using Human Chorionic Gonadotropin (HCG) combined with a low calorie diet of about 500 calories per day. HCG will be taken in the form of injections for a period of 23 to 42 days while observing a strict, food specific, very-low-calorie-diet (VLCD). While on the HCG Diet, the body begins to use its excess fat stores for fuel. It is theorized that the hypothalamus gland, the master gland in the brain that controls appetite and fat burning is "reset" leaving you with a new metabolic set-point. It has been said that the HCG helps to release somewhere between 1500-2500 calories per day for your body to use as fuel. This explains why you are not calorically deprived on the protocol.

After using the HCG and following the VLCD for 3 weeks or more, you will begin the next phase known as the "stabilization phase." In this phase, you will resume a normal calorie diet with a large variety of foods with the exception of starches and sugars for three weeks. Finally, in the "maintenance phase" you will begin to slowly incorporate carbohydrates back into your diet and enjoy your new body. You will have incorporated new dietary and lifestyle habits into your daily routine.

When was HCG and the HCG Diet discovered as a weight loss tool?

The HCG diet protocol was developed by Dr. Simeons in the 1950's. Through research, he discovered the connection between obesity and HCG. He found HCG to be extremely effective in correcting metabolic disorders by resetting the hypothalamus gland of the brain when his patients were also put on a specific diet. His patients had incredible results and improved metabolism over time. **We strongly recommend reading a free downloadable version of Dr. Simeon's protocol called "*pounds and inches.*"** Google "Dr. Simeons pounds and inches" to find it online.

How much weight will I lose?

If the program is followed correctly, you can expect to lose on average of .5 lbs – 1 lb. per day (some will lose more per day in the first week). Men tend to lose a little faster than women. The amount you lose also depends upon how much you have to lose and how strictly you follow the diet.

What is HCG? How is it thought to work with weight loss?

HCG is a peptide naturally produced in large amounts by the placenta during early pregnancy. It has many functions in utero and is used medically to treat a variety of conditions. **In the 1950's, it was discovered that the HCG Diet along with a rx HCG injection of a certain dosage had a special property that triggered the body to mobilize fat and use it for energy.** HCG was thought by Dr. Simeons to work on the hypothalamus of the brain where it encourages the body to release "fixed" or "abnormal" fat stores for fuel. When a person becomes overweight, there can be hormonal changes that make these "abnormal fat" deposits more challenging for the body to use as fuel. For many patients, they have found it extremely difficult to lose weight. By doing body composition analysis on our patients, we can see that the HCG Diet burns fat stores preferentially and spares skeletal muscle. This is extremely important in any diet program as "muscle dictates metabolism."

In order for the HCG Diet to be effective in burning stored fat, the specific HCG diet must be followed. The fat that is being mobilized by the HCG will only be burned if the specific low calorie diet is followed. As the fat is mobilized and used for fuel, hunger decreases in spite of the reduced intake of food. **Although you are eating a low calorie diet, the HCG is theorized to allow your body to burn on average a ½ to 1 lb. of fat per day.** Your body is not in starvation mode during this time.

Are there any clinical studies using this method of weight reduction?

Yes, there are quite a few. The HCG Diet has been in use for medical weight loss for over 60 years and had 16 years of clinical trials. Here are a few:

- http:/tinyurl.com/hcgstudy2
- http:/tinyurl.com/hcgstudy3
- http:/tinyurl.com/hcgstudy4
- you can continue with this link all the way through hcgstudy21

Does the HCG Diet have any side effects or contraindications?

Most of the HCG Diet side effects are positive ... we have seen brittle fingernails begin to strengthen, blood pressure and blood sugar normalize; cholesterol levels lower (but can be harmlessly higher during treatment); arthritis and rheumatoid symptoms will often improve. Other conditions that we have seen improve with the HCG Diet protocol are colitis, allergies, varicose ulcers, duodenal ulcers, gastric ulcers, psoriasis and migraines.

There are a few conditions in which we take precaution with the HCG Diet. One of them is a pre-disposition to gall stones. The diet itself could aggravate this problem because of the small amount of fat in the 2nd phase. Gout symptoms may also become more severe in rare cases. In rare cases, ovarian cysts may grow. In doing this protocol with over 1600 people, we have seen 3 patients have gall bladders removed; 2 patients stop due to ovarian cyst growth; 1 patient stop due to pregnancy induced goiter growth and one case of gout aggravation.

With any drug, there is always the possibility of an allergic reaction, yet we have not heard of any severe adverse reactions to HCG at these very low doses (200 -300 IU). Some may experience headaches in the first few days and this is most likely due to detoxification.

In huge doses (5000-10,000 IU), which are used for fertility treatments, side effects have been known and are listed in this link: http://tinyurl.com/hcgwarnings You can see drug interactions at this link: http://tinyurl.com/hcginteraction.

The side effects revealed by many patients are: decreased depression, increased libido, discovery of a new wardrobe, better lab results, more energy, lower blood pressure, stable blood sugar, ability to discontinue meds, optimistic outlook and a feeling of self-determination and control.

Is the Natural HCG Diet safe?

The Natural HCG Diet has proven to be very safe in our clinical experience. Women experience extremely high levels of HCG during the early months of pregnancy and can be pregnant many times without adverse effects. The small amount used for weight loss has had little to no side effects in our patient population over the past 10 yrs.

If the HCG Diet works for weight loss, wouldn't pregnant women lose weight?
HCG works to mobilize fat for utilization by the body when the HCG Diet food protocol is followed. If a pregnant woman consumed very little food through her pregnancy, she has the potential to give birth to a low/normal birth weight baby due to the fat mobilizing effects of HCG. For weight loss, we use a low calorie diet with rx HCG injections to help rid the body of fat. On body composition analysis before and after the program, we see that most of the weight loss comes from fat stores and lean mass stays consistent within a few lbs.

Wouldn't I lose the same amount of weight eating a very low calorie diet without HCG? Will my metabolism slow down?
If you just follow the very low calorie diet in phase II of this program and do not use the rx HCG, you will lose weight. The problem is, you will be very hungry, tired and most importantly you will be losing muscle and structural fat that we need for protection. Without HCG in combination with the specific diet, you will most likely not lose any of the excess fat deposits in the problem areas. In our clinical experience, we have seen the HCG Diet allow the body composition to change for the better and keep metabolism strong. We conduct body composition analyses on our patients to be sure little lean mass is lost.

The HCG Diet is very low calorie, will I get hungry? Is 500 calories per day safe?
Because the HCG Diet mobilizes fat and makes it available to the body as an energy source, it naturally reduces appetite in most people. Therefore, even though you are taking in fewer calories, your body is able to access caloric energy from your stored fat. Most people have plenty of energy and feel great while on the program. Sometimes, people report feeling hungry during the first few days. Loading properly in Phase I or the "loading phase" will help to alleviate this issue. We medically supervise our HCG Diet program to help trouble shoot any issues as they arise.

Research now confirms that cultures who naturally consume a calorie restricted diet actually have increased overall health and life span. When you are following this diet with HCG injections, it is hypothesized that your body is being flooded with about 1500 calories of energy from your excess fat reserves. This is why in phase II most people are not hungry and have increased energy levels.

Will the HCG Diet interfere with any medications I am currently taking?
The HCG at the dose given for the HCG Diet has not shown to interact with any medications. We have found that patients taking prednisone at higher doses lose weight at a much slower rate.

Will I experience any changes in my menstrual cycle on the HCG Diet?
Most women continue HCG through the menses with no problem. Occasionally, the menses may come a few days sooner or later than expected. Some woman may have increased or decreased bleeding while on the HCG Diet.

How much weight can I expect to lose on the program?
On average, our patients lose around 10 - 20 lbs. per 23 day course of HCG.
Of course, the amount of weight you will lose is dependent upon the amount of weight that you have on your body to lose. In the 42 day protocol, patients generally lose 20-36 lbs. Results vary greatly from person to person.

Does the weight loss slow down after the first few weeks?

Often times, there can be substantial weight loss during the first couple of weeks followed by a 3-4 day plateau. This does not mean weight loss has stopped. This is just the body's natural response to rapid weight loss. In most cases, you are still burning fat and simply holding water. The weight loss will resume again.

What are the 4 phases to the Natural HCG protocol in a "nut shell"?

- **Phase I** – "loading phase" (2 days) ~ during this phase you will follow the recommendations for loading your body with lots of healthy fats while beginning the HCG injections. It will take 48-72 hours for the HCG to begin mobilizing fat in your body as you begin to follow phase I for 2 days and then the specific HCG Diet in Phase II.
- **Phase II** – "HCG phase" (21 or 40 days) ~ the HCG is given daily and the food-specific, very low calorie diet (VLCD) is followed. The last 2 days of this phase are a "transition phase" where you will follow the VLCD without the use of HCG. You will weigh yourself daily and measure yourself weekly throughout this phase.
- **Phase III** – "stabilization phase" (21 days) ~ this phase begins 2 days after your last dose of HCG. This is IMPORTANT as this is the part of the program that stabilizes your weight at a new set point. During this phase, you are to avoid all sugars and starches and weigh yourself daily to be sure you are within 2-4 lbs. of your last injection day weight. If you go above 2-4 lbs., there are methods to be followed.
- **Phase IV** – "the maintenance phase" (ongoing) ~ at this stage, you can begin to add in small amounts of healthy starches and sugars very gradually and begin to follow a "normal," healthy diet. It is still important to weigh yourself each morning to track your weight and be sure your eating habits remain in balance. We will coach you on a way to eat that does not feel deprivational and will assist with maintaining your weight and feeling great.

What if I cheat?

If you cheat, your caloric intake could suppress your stored fat from being mobilized. **On this program, it is truly IMPORTANT that you do NOT CHEAT as you may sabotage your hard earned progress.** If cheating happens, recommit and jump back on board with the plan right away!

Is it important to use oil-free cosmetics and creams during this program?

Yes, it is best to use oil-free lotions! In the HCG DIET TOOLS section of this manual is a list of body care products that will work with this program. The oils in lotions, creams and ointments are absorbed through the skin and can interfere with weight reduction in some cases. Dr. Simeons found that those who contacted oils in their profession while doing the HCG program did not lose weight until they began wearing gloves or stopped skin contact with oils. Mineral oil particles are too big to be absorbed by the skin and body lotions in a base of mineral oil seem to work just fine with the HCG Diet Protocol.

Is it important to eat organic food?

Yes! I believe it is important to have clean sources of food. If you continue to ingest chemical pesticides, toxins, growth hormones and antibiotics you will gain the weight back over a couple of years. These toxins "clog" our systems. Please make eating organic a priority! Google "dirty dozen and clean 15" to learn what fruits and veggies are the most heavily sprayed.

Can HCG cause cancer?

NO! HCG is a cancer marker for gestational trophoblastic tumors and some germ cell cancers. This DOES NOT mean that HCG in your system is a bad thing. Otherwise, all pregnant women would be afraid their pregnancy would give them cancer. The truth is that HCG has been used to TREAT cancer. http://tinyurl.com/cancerstudy1 http://tinyurl.cancerstudy2 http://tinyurl.com/cancerstudy3 These links show HCG as being effective in treating breast and prostate cancers. You can type in the link http://tinyurl/cancerstudy and add the number 1-20 to see the different studies.

What are the HCG delivery methods?

With our program, HCG is taken into the body using subcutaneous injections via a thin insulin needle. The injection is given just below the skin. Injections are often more effective and is the preferred method of administration. We will provide you with pre-filled syringes each week that you are on the "Natural Path HCG Diet Program."

*the FDA does not approve the use of HCG for weight loss. This is an off label use of the medication.

PREPARATION PHASE
&
PHASE I "THE LOADING PHASE"

CONGRATULATIONS!

your body & your life are about to change ☺

PREPARATION FOR THE DIET

STEP 1 READ THIS MANUAL IN ITS ENTIRETY AT LEAST TWO TIMES

STEP 2 READ DR. SIMEONS "POUNDS AND INCHES" ALL THE WAY THROUGH
Google: "Dr. Simeons pounds and inches"

STEP 3 BE READY FOR THIS CHALLENGE. Commit to sticking to the program. Think about different daily menu plans. Be inspired!

STEP 4 PREPARE LOGISTICALLY BY ACQUIRING WHAT YOU WILL NEED.
A list of products can be found in the HCG TOOLS section. (p. 34 & 35)

- Food scale (digital that measures in ounces)
- Bathroom scale (digital that measures in .2 lbs.)
- Tape measure
- Food items (shopping list on page 34 & 35)
- Proper body care products (oil free or mineral oil based body lotion)
- Supplements if recommended by the doctor

STEP 5 WEIGH, MEASURE and PHOTOGRAPH yourself before you begin.

RECOMMENDED SUPPLEMENTATION ON THE HCG DIET (optional)

One or more of these supplements may be recommended while on our HCG diet program:

- **ULTRA BALANCE DF PROBIOTICS** ~ 1 capsule taken first thing in the morning away from food. This will help with cleansing and correcting the balance of flora in your GI tract. A healthy gut leads to a healthier you. *Did you know that the "good gut bugs" (probiotics) make more serotonin than the brain and that they are part of our primary immune system. They are also key players in how we digest, absorb and assimilate most of the food that we eat.*

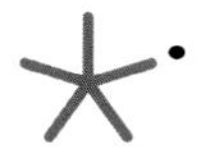

- **CANDIBACTIN –** 2 tablets in the evening to eradicate candida, yeast, fungus and bacteria. Decreases sugar cravings and intestinal inflammation. Promotes detoxification and can improve weight loss. *"Bad gut bugs" produce toxins that trigger inflammation and insulin resistance or pre-diabetes and thus promote weight gain. They can send a chemical signal to our brain to tell us to crave sugar as that is what they like to eat.*

We find our patients with sugar cravings do better with the candibactin/ultra balance probiotic combination.

- **SLIM SHOT Rx** ~ We highly recommend our patients have a weekly "slim shot" injection to help them with energy, fat break down, metabolism, liver function and immunity during phase II and phase III of the HCG Diet. This injection is optional and is $40 (see page 24)
- **SLIM DOWN** ~ An energizing formulation that is a nice adjunct to the HCG Diet. Helps with energy, improves fat burning, curbs appetite and helps keep blood sugar stable to diminish carb cravings. * Not for those with elevated blood pressure.
- **POTASSIUM** ~ 1 capsule before bed; potassium can be lost in the urine and it is necessary to replace it to keep your muscles feeling strong and your body in balance
- **CALM ~** a very gentle laxative that also calms nerves and promotes restful sleep
- **GLUTAMINE –** 2 capsules opened and placed under tongue to help with food/sugar cravings

Welcome to ... **THE LOADING PHASE**

PHASE I ~ FAT LOADING (DAYS 1 & 2)

Welcome to the first 2 days of this HCG diet program!
On these 2 days, you will begin by giving yourself your HCG injections.

THIS IS YOUR DAY TO EAT A VARIETY OF HEALTHY, HIGH FAT FOODS

Phase I is a very important part of the program. We find that patients who "skimp" on loading may experience more hunger during the first few days of phase II. In this process, you are letting the HCG build up in your blood stream while filling your fat reserves. Doing this phase well makes the transition into phase II much easier. We recommend choosing **clean, healthy, organic fats**:

- Higher fat proteins such as bacon, salmon, sausage or leaner meats with lots of butter or coconut oil (organic, grass fed or wild caught preferable)
- Veggies with butter, coconut oil, olive oil or tahini
- Avocado's and olives
- Nuts and nut butters
- High fat organic dairy products such as full fat yogurt, heavy whipping cream or sour cream

You may also indulge in other food and drink. However, we recommend having limited amounts of foods high in sugars and carbs.

Enjoy! Indulge in your favorite healthy, high fat foods. Eat until you are comfortably full and not overly full. It is not unusual to gain 1-3 pounds during this time (this comes off quickly when the diet begins).

IMPORTANT: The loading phase is a very important part of the program & must be done correctly. Those who do not load properly often feel hungry during the first week!! If you have dietary restrictions we can help you prepare an alternative loading menu.

Sample Loading Menu

Breakfast: Full fat Greek yogurt
Snack: ½ c. cashews; ½ c. macadamia nuts; avocado
Lunch: Sausage or high fat organic meat, vegetables in olive or coconut oil; olives
Snack: Organic ice cream or COCONUT milk based ice cream if you need a treat
Dinner: 20% fat organic ground beef; salad with tahini, avocado and walnuts.
Late night snack: handful of organic nuts

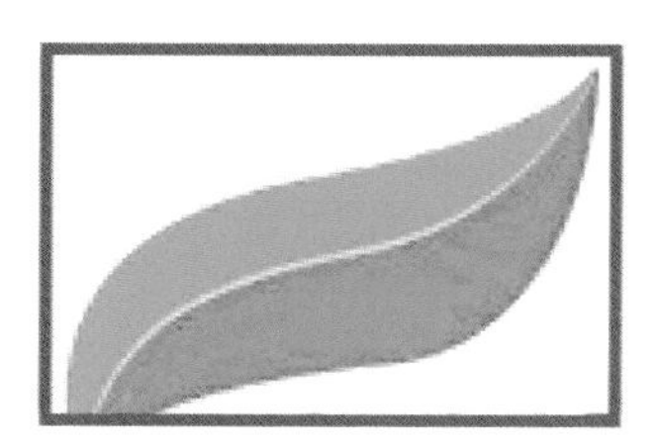

HCG PROGRAM

PHASE II ~ THE HCG PHASE

DAYS 3-23 or DAYS 3-42

PHASE II ~ THE HCG PHASE (Day 3 - Day 23 and can go on for up to 49 days)

THE 500 CALORIE DIET OVERVIEW

This portion of the HCG program must be followed exactly as outlined in order for optimal results to be seen. NO CHEATING! **There is a synergy to the foods allowed on Phase II that lead to the amazing results.**

The following is a basic outline of what can be eaten throughout the day. The menu can be broken into smaller, more frequent meals if needed. If you are a breakfast person, you may move some of the lunch or dinner protein and some of the allowed fruit to the morning time.

MORNING TIME~

1. Tea or coffee in any quantity; no cream or sugar. Stevia is the only sweetener allowed.
2. OPTIONAL ~ a nice way to begin the day is by combining 3 oz. water, 1 tbsp. lemon juice and 1 tbsp. apple cider vinegar. This will assist with cleansing & improve weight loss.
3. Drink 16 ounces of water

*egg whites are an optional breakfast food. However, in some cases the use of egg whites can slow down weight loss. If you hit a plateau, remove the egg whites and see if you begin to lose weight again.

LUNCH & DINNER (please choose **different** protein and vegetable for lunch & dinner if possible)

1. 3.5 ounces of ORGANIC **beef, chicken breast, white fish, lean ground turkey, lobster, scallops, water packed tuna, buffalo, crab or shrimp**. All visible fat must be trimmed away and the meat weighed when it is raw. It must be boiled or grilled without additional fat. The chicken breast must be boneless/skinless.

2. One large handful (1-2 cups) of an ORGANIC vegetable from the following list: **spinach, chard, chicory, beet-greens, green salad (lettuce or arugula), broccoli, cauliflower, red/green bell peppers, tomatoes, kale, collard greens, dandelion greens, celery, fennel, onions, radishes, cucumbers, asparagus, mushrooms or cabbage.**

3. An ORGANIC **apple, orange, a handful of strawberries or one half grapefruit** (can also be used as a snack or with your breakfast if you choose not to have this as part of your lunch.) Fruit can actually be optional and if you leave out the fruit you can have an extra 2 servings of vegetable.

The above menu contains the foods that you can have for **one day** during the HCG phase. There is no objection to breaking up the 2 meals in a way that works best for YOUR body!

Phase I is the "loading phase" – this is for the first 2-3 days of the program. On this phase, you begin your rx HCG injections and focus on "loading" up on healthy, high fat foods. There are no calorie or food restrictions on these days.

HCG DIET PHASE II FOOD and CALORIE CONTENT

2 servings of 3.5 ounces protein/day; 2 servings of 1-2 cups vegetable/day; 1-2 fruits/day

ORGANIC BEEF calories/oz. (3.5 ounce/serving)

- *Lean ground beef 37 calories
- Sirloin tip steak 52 calories
- Top round steak 56 calories
- Lean Buffalo 31 calories

ORGANIC POULTRY calories/oz. (3.5 ounce/serving)

- Chicken breast 27 calories
- *Lean ground turkey 32 calories

SEAFOOD calories/oz (3.5 ounce/serving) (up to 4 oz. white fish)

- Cod 27 calories
- Crab meat 31 calories
- Flounder/Sole 33 calories
- Haddock 32 calories
- Halibut 40 calories
- Lobster 29 calories
- Red snapper 25 calories
- Shrimp 34 calories
- Tilapia 36 calories
- Scallops 26 calories
- *Tuna (water pack) 27 calories

VEGETABLES calories/oz. (1-2 cups/serving; measure raw) (can mix 1-3 veg. together if not on a plateau)

- Asparagus 6 calories
- Celery 4 calories
- Cabbage 7 calories
- Cucumber 3 calories
- Spinach 7 calories
- Radishes 5 calories
- Lettuces 5 calories
- Chard 7 calories
- Fennel 9 calories
- Onion 12 calories
- *Broccoli 10 calories
- *Cauliflower 14 calories
- *Mushrooms 7 calories
- *Bell Pepper 6 calories
- *Kale 14 calories
- *Collard greens 7 calories
- *Tomato 5 calories

Free Veggies

FRUIT calories/oz. (1-2 servings/day; can omit fruit options and add extra vegetable serving)

- Apple (medium) 15 calories
- *Orange (medium) 13 calories
- Grapefruit (half) 9 calories
- Strawberries (cup) 9 calories

PROTEIN ALTERNATIVES calories/oz

- 1 egg + 3 egg whites 78 cal./egg + 17 cal./egg white
- ½ cup low fat organic cottage cheese or 0% fat Greek yogurt 80 calories
- Protein Powder (Sun Warrior Vanilla, Warrior Food Vanilla, Thorne Vegalite Vanilla, Garden of Life Raw Protein Vanilla**) *** Garden of Life is in most New Seasons and Whole Foods stores*

Lunch: 3.5 ounces of allowed protein with 1-2 cups of allowed veg (break up into 2 small meals if desired)
Dinner: 3.5 ounces of allowed protein with 1-2 cups of allowed veg (break up into 2 small meals if desired)
Fruits: 2 servings of allowed fruit per day
Snacks: chopped celery, cucumber, cabbage, radish, etc.
Water: 2-4 L/d

If a **PLATEAU** happens: please omit the foods with a * after them. Please omit fruit or have just one serving/day. You can replace them with a veggie serving. Do not mix vegetables. Be sure to have a minimum of 2L water/tea daily. Be sure you are not constipated – please use a gentle laxative when needed. Try a morning walk or gentle exercise in am. If the plateau is more than 3 days -- go for an "apple day." Plateaus are NORMAL! Do not worry☺

PHASE II IMPORTANT REMINDERS & TIPS

1. The juice of one or two fresh lemons per day is allowed for all purposes.
2. Salt, pepper, distilled vinegar, raw apple cider vinegar (no balsamic vinegar), and most fresh or dried herbs may be used to season. Read all labels carefully!
3. NO oil, butter or dressings (unless HCG Diet approved dressing w/out oil).
4. Tea, coffee, water & mineral water are the only drinks allowed. Stevia may be used to sweeten your beverages. Sweet Leaf makes many flavored stevia drops. Great in sparkling water if you need a sweet treat. (sweetleaf.com)
5. Drink at least 2-4 liters of pure water each day
6. It is okay to mix 1-3 vegetables together (undo if plateau happens)
7. Avoid oil-based body lotions. The skin has the ability to absorb the oils and utilize them as a fat source. Mineral oil based products or oil free are the best.
8. Weigh yourself DAILY upon rising after using bathroom (if needed).
9. Be sure to weigh your protein in the RAW form (100 grams or 3.5 oz.)
10. For women, the menstrual period can cause a plateau. This will shift.
11. Consider a daily dose of light exercise ~ 15-30 minutes per day.
12. Try to get plenty of sleep ~ bed by 10 pm is a nice way to go.
13. **Take a cleansing bath as many evenings as you can.** Use 2 cups epsom salts & 1 cup baking soda and relax in water as hot as you can take for 15-20 minutes.
14. The food on the phase II food list can be eaten anywhere you would like throughout the day. Please find a rhythm that works best for you!
15. Check out **HCG Diet Cookbooks** and the internet for great recipe ideas!
16. Always stay prepared. Have food on hand for when hunger strikes!
17. Keep busy. Read a book. Go for a walk. Boredom can lead to mindless eating.

NOTE: Remember that your body is going through a deep cleansing process as your body is mobilizing stored fat. You may feel tired or sluggish for the first few days. It will pass. Be gentle with yourself. The way to your new body is through these uncomfortable moments if they should arise.

CHEATING

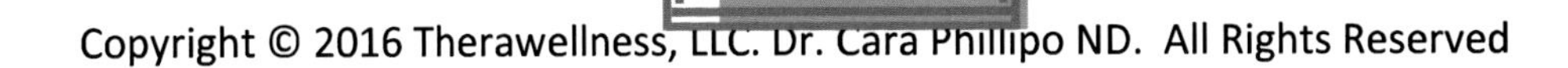

INTERRUPTION OF WEIGHT LOSS: PLATEAU and PLATEAU BREAKERS

A plateau is a NORMAL part of any rapid weight loss program. A plateau lasts 3-4 days and frequently occurs about half way through Phase II/ HCG Phase of the program. It is not uncommon for a menstruating woman to have an interruption of weight loss a few days before and during the menstrual period.

Remember: a plateau always corrects itself!

When a plateau occurs and lasts more than 2-3 days we recommend that you try an apple day (see below). If you hit a plateau, please feel free to contact Dr. Madda & Dr. Phillipo to help you to get back on track with weight loss. Sometimes, plateaus cannot be explained and weight loss will resume when the body is ready. The body can hold water for many various reasons.

During a plateau we recommend that you:

- Omit broccoli, cauliflower, bell pepper, mushrooms, tomato, orange and beef
- Eat only one vegetable per meal (no more mixing)
- Walk for 20-30 minutes in the morning
- Drink plenty of water and tea
- Take a gentle laxative if bowels are not moving
- After 3 days of a plateau we recommend the APPLE DAY

Plateau Mantra:

"I am still burning fat; I am simply holding water and this too shall pass."

AN APPLE DAY: an apple day begins at lunch one day and continues until lunch the following day. You are allowed to eat up to 6 small sized apples during this 24 hour period (one full day until lunch the following day). Drink plenty of water and herbal teas.

The Lipotropic "Slim Shot" Injection

We highly recommend our patients have a weekly "slim shot" injection to help them with energy, fat break down, metabolism, liver function and immunity during phase II and phase III of the HCG Diet. This injection is optional and is $40.

Assist your outcome with the weekly "Slim Shot" Injection

- Boosts the breakdown and metabolism of fat
- Helps metabolism on a cellular level
- Increases energy
- Assist mental clarity and focus

Lipotropic nutrients have been known to assist with lowering appetite and increasing your body's natural fat-burning processes. Using lipotropic nutrients, along with proper diet and exercise can help you to reach your weight management goals.

This injection consists of a mixture of amino acids and vitamins. It is given in office once per week on phase II & phase III of the HCG Diet program. Many of our patients say that this injection improves energy, mental clarity and focus.

What are the ingredients in the "Slim Shot" injection on the HCG Diet

Methionine is an amino acid that acts as a lipotropic (fat loving) agent to speed up the removal of fat within the liver & to prevent excess fat buildup in problem areas. It helps to detoxify the body of heavy metals and is considered a strong anti-oxidant. Many report a boost in energy from Methionine.

Inositol exerts lipotropic effects as well. An "unofficial" member of the B vitamins, Inositol has also shown to help with break down and re-distribution of body fat.

Choline is essential for fat metabolism. Choline is known to assist our bodies to efficiently burn fat. Choline supports the health of the liver, improving its ability to process and excrete chemical byproducts within the body.

Vitamin B12 is a vital nutrient that is critical for maintaining proper functioning nerve cells; helps to produce DNA, and RNA; and boosts fat metabolism in the body. Vitamin B12 is also referred to as the "energy vitamin." Many patients report an energy boost after a Vitamin B12 injection.

B complex (vitamins B1, B-2, B-3. B-5 B-6, B-7, B-9) plays an important role in keeping our bodies running like well-oiled machines. These essential nutrients help convert our food into fuel, allowing us to stay energized throughout the day. Great nervous system support.

Carnitine - The primary function of carnitine in the body is to regulate fat oxidation (burning). L-Carnitine is responsible for transporting fat to the fat furnace in our cells called mitochondria. Unless fat makes it to the mitochondria, it cannot be oxidized.

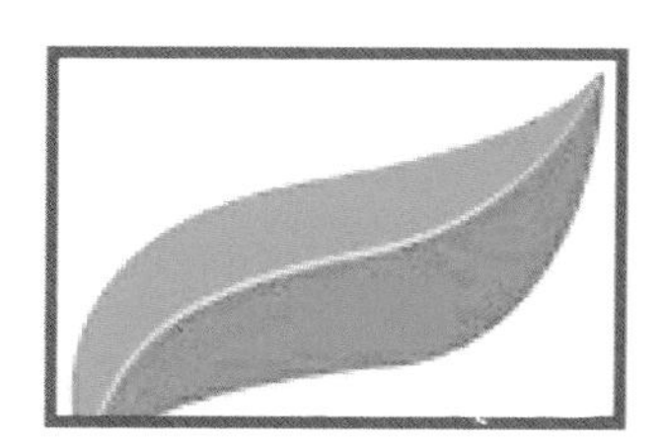

HCG PROGRAM

PHASE III~ THE STABILIZATION PHASE

(2 days of transition and then 21 days)

PHASE III ~ THE STABILIZATION PHASE of the Natural HCG Diet Program

If you have had great success with the HCG Diet, the last thing you want to do is let go of the incredible work that you put forth to create these amazing changes in your body. Be mindful and aware if you notice that you are resorting back to old eating habits.

PHASE III *- The first 3 weeks after phase II of the HCG Diet are* ***IMPORTANT*** *to maintaining your weight loss. We have seen this clinically in facilitating this program with over 1600 patients in the past 10 years!*

Please understand as you get ready for Phase 3:

- Your body will take some time to get used to this new weight as a "set point." Your body may want to replace some of those fat stores that you have just worked so hard to lose. Be careful! If you want to keep your weight off, you will need to make dietary changes in your life. Choose foods that NOURISH your body on a cellular level. Eat only when you are hungry and stop eating when satisfied.
- While on Phase III, begin to think about what your general dietary lifestyle will look like and make the choice to follow that 90% of your life. Let that other 10% be the time you go out to celebrate with friends and with food.
- Careful with Carbs!!! Carbs are the main reason that people gain weight. We can get a sufficient number of carbs daily from eating vegetables and the occasional fruit or grain. These days, we all know where our next meal is coming from and there is not a true need for our bodies to have large storages of carbs or fat. Highly refined carbs are rapidly digested and the rush of incoming fuel after consumption is not utilized most of the time (unless you are out running a marathon). These excess calories get stored as fat. Please eliminate all processed carbs and sugars from your diet. Be careful with alcohol as it is high in sugar and hard on the liver.

THE BASICS of PHASE 3

DAY OF LAST INJECTION AND 2 TRANSITION DAYS

- Record your LIW (last injection weight) and follow the phase II protocol for 72 hrs or 3 days after your last injection. These are called "transition days."
- If you get hungry on the 2nd day of this transition phase you may go for some additional vegetables or an extra 3 ounces of protein.

PHASE III "The Stabilization Phase" ~ 21 days

- In Phase III, eat a healthy diet and be sure to have **NO SUGAR, NO CARB & NO STARCH** outside of the low sugar fruit, low carb vegetable and protein realm. You can feel free to eat an unlimited amount of low carb vegetables!
- In phase III, you can have a conservative amount of fat as directed by your Dr.
- **Eat when your body feels hungry and stop eating when you are satisfied.** You will work up to eating 1200-1500 calories per day (2000 calories for some men).
- The GOAL of phase III is to maintain your new weight within a 2-3 lb range. This is important as we are setting the hypothalamus at this time with your new "base weight." A small amount of weight loss or gain may happen in this phase.

PHASE III SUGGESTIONS FOR SUCCESS

- Keep a food journal. This will help identify the foods that cause weight gain
- Eat no more than 1-2 servings of fruit per day. One serving would be ideal.
- Avoid high sugar fruits (cherries, kiwi, mango, pineapple, papaya)and starchy vegetables (corn, potato, beets, carrots, peas, squash)
- Eat nuts, seeds, organic cheeses, milk and yogurt sparingly; one serving per day
- Choose Organic foods always. It is the toxins in food that also contribute to weight gain and hormonal disruption.
- Have a nice protein source with each meal; keep carbs under 40 grams per day
- After dinner, enjoy a nice cup of tea and eat very lightly

WHEN PEOPLE HAVE A HARD TIME ON PHASE III IT IS USUALLY BECAUSE OF:

- Not eating enough food, **especially PROTEIN.** You should be eating a minimum of 1000 -1500 calories per day. Trying to eat too little to avoid gaining back the weight you have just lost is a common mistake. While on the HCG phase, you were releasing a large number of calories each day for your body to use for fuel. In addition, you were eating another 500 (ish). You can see how if you deprive your body of calories during phase III it may go into a "starvation" mode and want to store fuel as fat. A general rule is to practice mindful eating ... **EAT WHEN YOU FEEL HUNGRY AND STOP EATING WHEN YOU FEEL SATISFIED!**

IMPORTANT PHASE III INFORMATION:

- **PROTEIN is very important here** and it is important to have some with every meal. Protein deficiency is a common reason for water retention in the body. Eat the amount of protein that feels right for you.
- **You should never gain more than 3 lbs. without immediately correcting it.** If you go 3 lbs over your last injection weight (LIW) it is time for a "CORRECTION DAY."

- **You <u>MUST</u> weigh every morning** after using the bathroom to make sure that you have not gained more than 3 lbs. since your LIW. You can easily gain 5-7 lbs. without noticing it in your clothes. If you end up a few pounds over where you would like to be … a correction day is an easy, non-deprivational fix. This can be done any time in your life.

- <u>CORRECTION DAY</u>: If you gain 2-3 lbs. over your last injection weight. <u>DO SOMETHING ABOUT IT that very day</u>. Eat only phase II proteins and vegetables in phase 3 quantities (unlimited veggies and 4-7 oz protein/serving). Have one fruit this day. Have a piece of fruit for breakfast; a lean protein (4-7 oz.) and a non-starchy vegetable for lunch and the same for dinner. Munch on the HCG Diet free veggies as snacks and drink lots of water. Your weight should be down the next day.

HCG Diet Phase 3 / STABILIZATION PHASE Summary

- You begin this phase 72 hours after your last injection of HCG.
- You will gradually increase your calories from 500/day to 800/ day or two and then to 1000-1500.
- You can reintroduce many foods during phase 3 except sugars and starches.
    - <u>Sugars</u>: Anything with 4 grams or less
    - <u>Carbs:</u> Anything with 10 grams or less
- You can begin exercising moderately. Increase your exercise gradually.
- Weigh in every morning. Use a "CORRECTION DAY" if you go 2-3 lbs over LIW.

Phase 3 of the HCG diet is very important. During this phase you are training your body to accept a healthier nutritional lifestyle. It is extremely important to **remain committed!**

Start Phase 3 SLOWLY! After the HCG phase, do not fall into this trap of wanting to devour food. Start your P3 very slowly – the recommendation for the first couple of days is to concentrate on P2 foods, just increase the protein, mix your veggies and add minimal fats. You can have 1 fruit per day (recommended) or 2 max.

- **Track Your Progress.** Data is very important. Keeping a food journal is essential to maintaining your weight loss. This will not be forever. You will learn what effects some foods have on your weight and the way that you feel.

  Take Action Immediately! As soon as you gain 2 pounds over your LIW, take care of it. Procrastination is your ENEMY. Look into the different ways of doing "Correction Days".

A Note on Alcohol- Most alcoholic beverages have very high sugar content. During HCG Diet phase 3 you will want to avoid those beverages. If it has less than 4 grams of sugar per serving, try it and observe your weight.

Phase III Food List (1200 to 1500 calories/day for 3 wks) *not complete – general idea

PROTEINS:

All fish including:
Flounder
Herring
Salmon
Sardines
Sole
Tuna
Trout

All fowl including:
Cornish hen
Chicken
Duck
Turkey

All shellfish including:
Clams
Crab meat
Mussels*
Oysters*
Shrimp
Squid
*Oysters & mussels have higher carb – 4 oz./day

All meat including:
Lean Beef
Bacon*
Ham*
Lamb
Pork
Veal
Venison
*Some processed meat, bacon, and ham is cured with sugar, which will add to the carb count. -Avoid cold cuts and other meats with added nitrates.

VEGETABLES

Salad Vegetables:
Alfalfa sprouts
Arugula
Bok choy
Celery
Chicory greens
Chives
Cucumber
Endive
Escarole
Fennel
Iceberg lettuce
Mushrooms
Parsley
Peppers
Radishes
Romaine lettuce

Non Salad Vegetables:
Artichoke
Asparagus
Artichoke hearts
*Avocados
Broccoli
Brussels sprouts
Cabbage
Cauliflower
Swiss chard
Collard greens
Eggplant
Hearts of palm
Kale
Leeks
Okra
*Olives green & black
Onion
Rhubarb
Sauerkraut
Spaghetti squash (moderate)
Spinach
Summer squash
Tomato
Zucchini

**Olive and avo are also fat sources and should be eaten in moderation on P3

FRUITS

Blueberries (1 cup)
Raspberries (1 cup)
Strawberries (1 cup)
P2 fruits – apple, orange, grapefruit

FATS (Oils)

Butter (REAL not imitation or margarine)
Mayonnaise – make sure it has no added sugar
Olive oil
Vegetable Oils
Nut Butters (1-2 tbsp)
Nuts (small quantities; 10 nuts)
Grape seed oil
Sesame oil
Sunflower oil
Coconut oil
Avocado
Olives
Dairy

USE THE HUNGER SCALE & FULLNESS SCALE WITH EVERY MEAL ON PHASE III

HUNGER SCALE (best to eat at a level 2- 2.5)

Level 1 – SATISFIED ~ your body feels perfectly comfortable and there is no hunger

Level 2 – SLIGHT ~ your body has begun to notice the hunger sensation. Now is a good time to begin to think about your next meal. You can probably wait 10-15 minutes or so.

Level 3 – EAT NOW ~ at this point, your hunger level is uncomfortable and you should have eaten 10 minutes ago. Foods that are quick and easy become appealing.

Level 4 – URGENT ~ you are now feeling irritable, light headed and maybe even getting a headache. You could eat almost anything in large quantities.

FULLNESS SCALE (best to stop eating at a level 1)

Level 1 – SATISFIED ~ your body feels perfectly comfortable and you sense food in your belly

Level 2 – SATIATED ~ you have eaten a bit too much and you sense a bit of discomfort in your belly. You have no hunger here ... if you choose to eat it requires emotional justification (not physical need).

Level 3 – VERY FULL ~ you are very uncomfortable at this point and your belly feels bloated. You could easily wait 3 or more hours until your next meal.

Level 4 – PAIN ~ you are so full you now feel sick. You are having indigestion and would love to lie down.

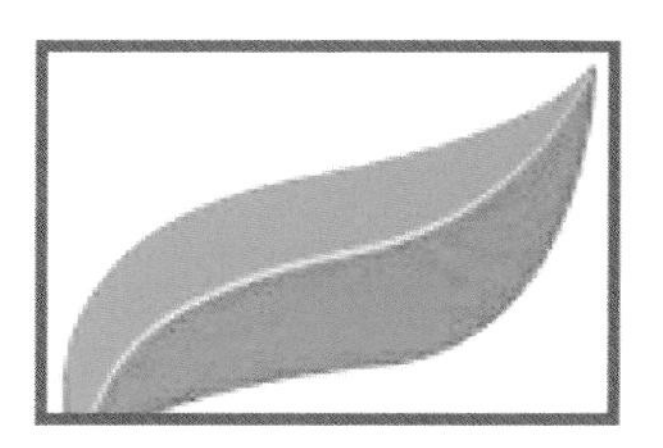

HCG PROGRAM
PHASE IV~ THE MAINTENANCE PHASE

PHASE IV ~ THE MAINTENANCE PHASE ~ *the rest of your life phase*

After three weeks of the stabilization phase, you will finally be at the maintenance phase. You have now supported and rebalanced your metabolism and it is time to very gradually add healthy starches into your diet with moderation if you desire. You could easily leave out starches other than the fruit and vegetable realm and continue a modified phase III diet and add in some additional healthy fats.

WE HIGHLY RECOMMEND FOLLOWING THE PALEO DIET FROM HERE 90% of the time. That means that 10% of the time you can have other foods that are not part of the paleo plan. This may be a couple dinners per week or even a "cheat afternoon" each week.

We do not recommend eating grains more than once or twice per week if needed. It is best to get your carbohydrates through non-starchy vegetables and low sugar fruits.

We recommend beginning your day with 30-40 grams of protein at breakfast time (when your body is hungry). This can be done with animal protein, eggs, Greek yogurt, cottage cheese or a clean source of protein powder. We recommend only light snacking of fruits and vegetables between meals and not eating past 7:30 pm. Lots of pure water and herbal teas are wonderful.

We recommend doing a morning weight at least once per week. Make sure you do not gain more than 3 lbs. from your last injection weight. Should a gain of over 3 lbs. occur, go back to doing the Phase III "correction day" for 2 days and this should bring your weight back down. This tool is always available to you!

Approach PHASE IV or the MAINTENANCE PHASE as a great adventure in finding out what you can eat and still maintain your weight. You will learn which foods cause you to gain weight and which ones do not. The food combination to watch out for is **fat and starch** together. This combination should be kept for the occasional indulgence.

Exercise every day!

Drink WATER!! Half your body weight in ounces per day.

No artificial sweeteners, MSG, high fructose corn syrup, food additives or fast food.

After completion of the HCG Diet protocol, you can feel proud of the weight you have lost and the patterns you have changed. If you have additional weight to lose, you can begin a second round of HCG in 6 to 8 weeks after your last injection. It is important to take a rest between rounds.

CELEBRATE!

Disclaimer HCG is a drug which has not been approved by the food and drug administration as safe and effective in the treatment of obesity or weight control.

FOLLOW THE HUNGER AND FULLNESS SCALE IN YOUR LIFE EVERYDAY!
IF YOU ONLY EAT WHEN YOU ARE HUNGRY AND STOP EATING WHEN YOU ARE FULL YOU WILL MOST LIKELY NEVER HAVE A WEIGHT ISSUE

HCG DIET TOOLS

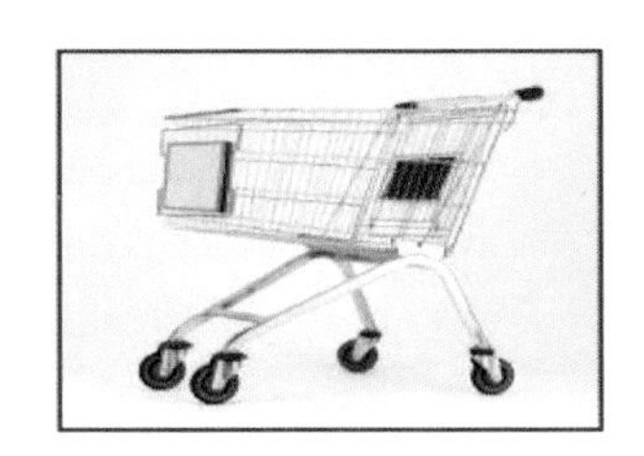

SHOPPING LIST FOR HCG PHASE

FOOD LIST

Organic Beef 3.5 ounces/100 grams

- Lean Ground Beef (93-98%)
- Cube Steak
- Sirloin Tip Steak
- Top Round Steak
- Buffalo (95%)

Organic Poultry 3.5 ounces /100 grams

- Chicken Breast (boneless/skinless)
- Lean Ground Turkey

Seafood 3.5 ounces/100 grams

- Cod
- Crab Meat
- Flounder
- Haddock
- Halibut
- Lobster
- Red Snapper
- Shrimp
- Tilapia
- Lemon Sole
- Scallops
- Water Packed Tuna (3 oz.)

Other Foods

- Chicken, Vegetable and Beef Broth ~ organic and low sodium; 1-2 cups/day
- Bragg's Liquid Amino's (use in moderation)
- Any seasoning or fresh herb that is 0 carb and 0 sugar (read labels)
- Liquid Stevia ~ comes in a variety of flavors
- Cinnamon, Celery Seed (great on salad), stone ground mustard

Beverages

- Mineral water
- Coffee/Tea
- Raw Apple Cider Vinegar
- Lemons (for lemon water)

Vegetables

- Asparagus
- Celery
- Cabbage
- Cucumber
- Lettuce all varieties
- Red Radishes
- Spinach
- Tomato
- Fennel
- Onion
- Collard Greens
- Cauliflower
- Kale
- Dandelion Greens
- Chard
- Broccoli
- Red/Green Bell Peppers
  Mushrooms

Fruit

- Apple (s) = 55 calories; (m) = 72 calories; (l) = 110 calories
- Orange
- Strawberries
- Grapefruit
- Lemon wedge

RECOMMENDED OIL-FREE BODY CARE PRODUCTS

Body Care

Jason anti-aging moisture cream
Jason Fragrance Free hand and body lotion SPF 15
Alba Hawaiian oil free moisturizer aloe & green tea
Lubriderm (mineral oil based) / Cetaphil
Mineral oil

HOUSEHOLD PRODUCTS

Measuring Tape
Digital bathroom scale
Digital food scale
Alcohol Pads for injections
Epsom salts (bath)
Baking Soda (bath)

HCG INJECTING

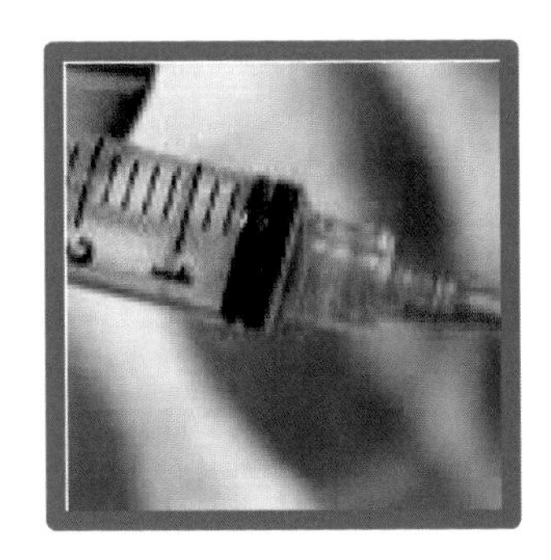

Giving yourself a **subcutaneous injection** means inserting the HCG into an area just under the skin. The needle used is small and thin and causes very little to no pain.

HCG is to be injected 1 time per day at the same time every morning (within a few hours)

INJECTING THE HCG

- **a. Areas: upper thigh or lower abdomen**
- **b.** Wash your hands.
- **c.** Choose your area and rotate site daily.
- **d.** Clean area with alcohol pad; 1 inch in diameter.
- **e.** Be sure the HCG in the syringe is free of color, debris and cloudiness.
- **f.** Insert the needle perpendicular to the skin; make sure the needle is all the way in; a quicker injection will create less pain.
- **g.** Continue with injection by slowly pushing the plunger down all the way
- **h.** Remove the needle from the skin and hold an alcohol pad at the injection site. Cap needle and place in sharps container or bag to return to office for proper disposal.

IF ANY STRANGE REACTION OCCURS PLEASE CALL US IMMEDIATELY 503-347-4625.

NEEDLE DISPOSAL: Needles should be disposed of in a safe manner and placed in a heavy duty puncture resistant container. Place both the needle and the syringe into the container. Keep this container out of the reach of children. Keep lid closed. Please return to the Natural Path and we will dispose of them properly. Thank-you.

Disclaimer: HCG is not approved by the federal Food and Drug Administration as a weight-loss medication. This is a legal but "off label" use.

HCG DIET PLAN

We are here to support you every step of the way on your HCG Diet journey. Our office phone is 503-347-4625. When we are not in the office, feel free to reach us by cell phone, text or email.

DR. CARA PHILLIPO Cell
415-306-6879
caraphillipo@gmail.com

DR. ROBERT MADDA Cell:
503-504-1772
robertmadda@gmail.com

WE CARE ABOUT YOU AND WANT YOU TO BE SUCCESSFUL WITH THIS PROGRAM!

Thank you for trusting us to take this weight loss journey alongside you!

It is an honor to have this opportunity to work with you!